ARTIFICIAL INTELLIGENCE: FRIEND OR FOE?

Artificial Intelligence in Transportation

Will AI Help Us or Hurt Us?

Nick Hunter

I0867438

CHERITON
CHILDREN'S BOOKS

Published in 2025 by **Cheriton Children's Books**
1 Bank Drive West, Shrewsbury, Shropshire, SY3 9DJ, UK

© 2025 Cheriton Children's Books

First Edition

Author: Nick Hunter
Designers: Paul Myerscough and Jessica Moon
Editor: Sarah Eason
Proofreader: Kate Hobson

Picture credits: Cover: Shutterstock/Metamorworks (foreground), Shutterstock/Suwin66 (background); Inside: Throughout: Shutterstock/Nadya Art, p1: Shutterstock/FrentaN p5b: Shutterstock/PopTika, p5t: Shutterstock/Heitor Pergher, p6: Shutterstock/Everett Collection, p7: Shutterstock/MikeDotta, p8: Shutterstock/Gorodenkoff, p9: Shutterstock/NadyGinzburg, p10: Shutterstock/Scharfsinn, p11: Shutterstock/Suwin66, p12: Shutterstock/Gorodenkoff, p13: Shutterstock/Metamorworks, p14: Shutterstock/Sundry Photography, p15: Shutterstock/Phil Pasquini, p16: Shutterstock/Hayk Shalunts, p17: Shutterstock/Gorodenkoff, p20: Shutterstock/Gorodenkoff, p21: Shutterstock/AlinStock, p22: Shutterstock/SariMe, p23: Shutterstock/Suwin66, p24: Shutterstock/Helloabc, p25: Shutterstock/Suwin66, p28: Shutterstock/Trong Nguyen, p29: Shutterstock/Gorloff-KV, p30: Shutterstock/Christopher Halloran, p31: Shutterstock/TierneyMJ, p32: Shutterstock/Aaron Kohr, p33: Shutterstock/Cono0430, p36: Shutterstock/Marko Aliaksandr, p37: Shutterstock/Skycolors, p38: Shutterstock/Aappp, p39: Shutterstock/Gorodenkoff, p43: Shutterstock/Piyaset, p44: Shutterstock/Gorodenkoff, p47: Shutterstock/Metamorworks.

All rights reserved. No part of this book may be reproduced in any form without permission from the publisher, except by a reviewer.

Printed in China

Please visit our website,
www.cheritonchildrensbooks.com
to see more of our high-quality books.

Contents

What Is Artificial Intelligence?

Artificial Intelligence (AI) is already supporting all of us whenever we travel. If you want to find a street address that you've never been to before, you may search the location via a mapping app on a computer or mobile device. These apps use AI to give us all the information we need for our journey: what's the best route? When will I arrive? Is there traffic on the journey? A navigation app uses mapping data and other data such as details about how much traffic is in a particular area. It seems to be thinking for us, but its use of data means it can give us useful information instantly.

HOW DO WE DEFINE AI?

Before we look at the many changes that AI could bring to transportation, let's consider what we mean by artificial intelligence. Artificial intelligence describes the way that software or machines can be designed and programmed to do things that can normally only be carried out by intelligent beings, including humans. These things could include recognizing human speech, making decisions based on data, and carrying out complex tasks based on these decisions.

NAVIGATION

We don't think greatly about how we move around, but even by just walking around your home you are using data and making complex decisions. First, your brain and your body combine to make you walk in a straight line without falling over. Next, your eyes take in data about the layout of the house and where every obstacle is so you don't bump into anything. You have a detailed three-dimensional (3-D) image of your house stored in your brain so you know where you want to go. Then, you may also respond to other real-time data such as the smell of food cooking, which tells you dinner is almost ready.

COMPUTER NAVIGATION

For a computer to **navigate** a building or cross a busy street, which seems natural to us, very complex processes and huge amounts of data are required.

It's tough enough for humans to navigate city streets—could a computer really do it better?

All that information is needed for a computer to respond to situations as we would. However, when very advanced computers master these complicated processes, it has a powerful impact on the world.

HUGE CHANGES

Since the beginning of the twenty-first century, there have been huge steps forward in the development of AI. Many people believe that the changes brought by this technology are just beginning and will affect all areas of our lives. AI will bring major changes to the way we travel and transport goods. This will bring benefits but could also bring risks and problems for many people. Some people are excited about how AI could change our transportation, others are more concerned about it.

This book will look at how AI could change transportation and **debate** whether this new technology is our friend or **foe**. Look for **IS AI A FRIEND OR FOE?** throughout the book. Read the arguments for and against AI, then answer questions that invite you to draw your own conclusions about whether this transformative technology will help us or hurt us.

CHAPTER 1

The History of AI in Transportation

Transportation has seen many dramatic changes in the twentieth and twenty-first centuries. One hundred and fifty years ago, most everyday transportation was powered by horses. Railroads and then motor vehicles brought a huge change in human travel. Since then, air travel has enabled us to travel to the other side of the world in a single day. AI may be the next transportation revolution, but some aspects of AI in transportation are not entirely new.

MACHINES THAT NAVIGATE

Machine learning is one of the key foundations of AI. Machine learning gives computers the ability to learn from data or from their surroundings without being specifically programmed. One of the first developments in machine learning was made in 1950 by Claude Shannon at Bell Laboratories. Shannon created Theseus, a mouse that could navigate its own way out of a maze. By today's standards, Theseus was very **primitive**, but it was one of the first examples of a machine learning from its surroundings. If its whiskers touched an obstacle, Theseus would change direction.

AUTOPILOTS AND GYROSCOPES

The first **autonomous** steering systems were actually created for aircraft. The Mechanical Mike

The first airplane **autopilots** were invented at the time of **pioneers** such as Amelia Earhart, who rose to fame as a pilot in the 1930s.

Some city trams and transit systems operate without drivers.

aircraft autopilot was created as early as 1933, just a few years after pioneers such as Charles Lindbergh and Amelia Earhart started to make long flights across the ocean. The autopilot was based on a **gyroscope** that kept track of the aircraft's course and was linked to the controls. Gyroscopes are an important part of all self-driving or autonomous vehicles. Aircraft autopilots and instruments have become very complex since these first examples and play a vital role in flying modern aircraft.

KEEPING ON TRACK

Autonomous vehicles have been common in the world's cities for many years, including rapid-transit systems, metro trains, or airport monorails. These forms of transportation may not have drivers on board. They usually run on rails so they don't have to make complex moves or deal with **unpredictable** traffic. They simply must be able to start and stop, open or close doors, or react to a hazard such as something unexpected on the track.

THE FIRST AUTONOMOUS VEHICLES

Self-driving vehicles have been imagined in science fiction stories and movies for many decades. The Space Race of the 1960s led to the development of the first basic autonomous vehicles. Researchers were attempting to solve the problem of controlling a vehicle on the moon. By doing so, they invented a vehicle named the Cart that was programmed to detect and follow a white line on the ground.

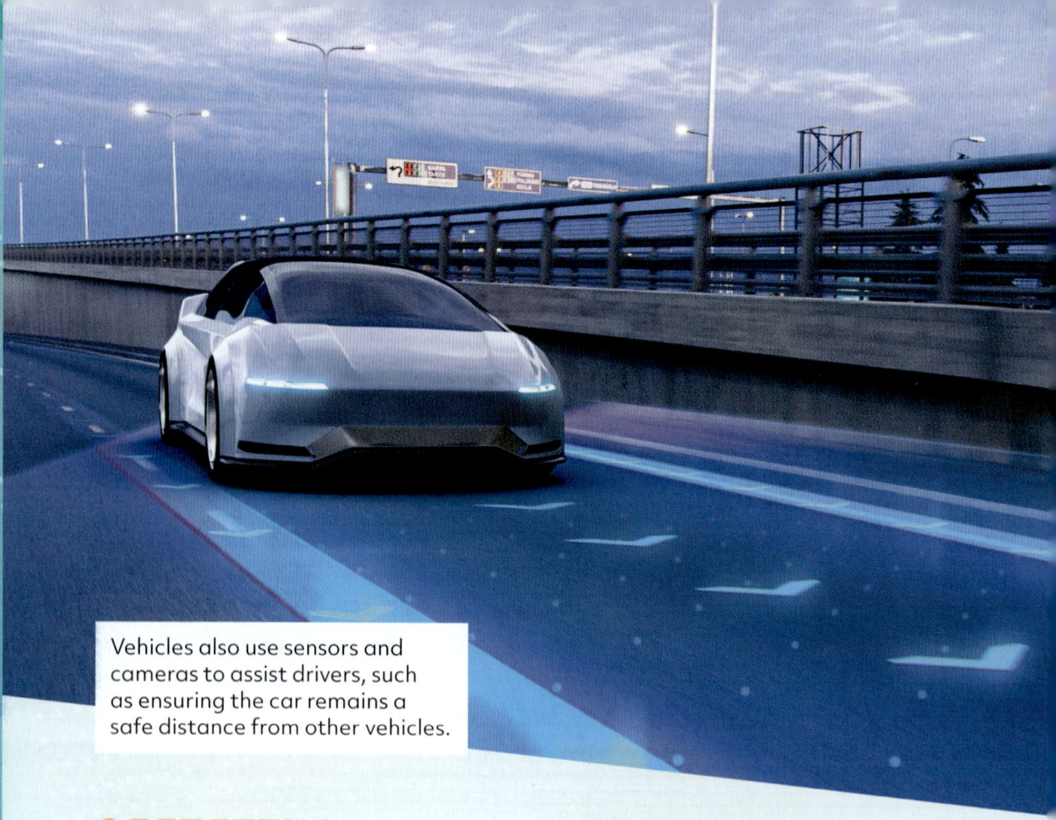

Vehicles also use sensors and cameras to assist drivers, such as ensuring the car remains a safe distance from other vehicles.

ASSISTING DRIVERS

Various experimental vehicles followed the invention of Cart, which could all recognize and follow road markings. However, they were still a long way from navigating on real roads, alongside other vehicles and **pedestrians**. Since the 1990s, there have been many new technologies that have assisted drivers, for example, by keeping a safe distance from the car in front or keeping on the correct side of the road. These technologies use elements of AI as they make decisions based on data provided by sensors and cameras. However, the technology is still quite limited and the driver is always in control.

TOWARD SELF-DRIVING CARS

The early 2000s saw a series of challenges to design truly autonomous cars that could navigate on roads while avoiding obstacles. Designers of these vehicles made rapid progress using the AI technology available at the time. In 2015, this self-driving tech hit the road when Tesla released an autopilot system for its cars. The technology allowed drivers to relax in the car while the self-driving vehicle did the driving—steering, changing lanes, and even parking as needed. Today, fully self-driving cars have covered millions of miles on real roads in testing. The self-driving revolution is truly here.

ENVIRONMENTAL IMPACT

As motor vehicles powered by **fossil fuels** took over the world, their inventors did not **predict** that billions of automobiles would create one of the biggest problems humankind has ever had to deal with—**climate change**. **Carbon dioxide emissions** from burning fossil fuels are causing our planet to heat up, and emissions from transportation make up around one-fifth of all carbon dioxide emissions. AI can have an impact on cutting these emissions.

Making Maps

It is no accident that the tech companies that pioneered mapping **software** are also involved in developing self-driving cars. In many ways, the two go hand in hand. The data from mapping software provided by companies such as Google and Apple is used to guide self-driving cars, although they need much more detailed and precise data than you would need to find an address. Every time these cars drive, they are adding to that data, updating the mapping software, and providing information for training other vehicles.

Driverless vehicles are powered by electricity so will cut vehicle **pollution**.

CHAPTER 2

The Self-Driving Revolution

Self-driving cars have been the future for some time. Tech billionaire Elon Musk predicted that there would be 1 million self-driving taxis on city streets by 2020. Although this didn't happen, you can now hail a **robotaxi** in some cities. Companies such as Google-owned Waymo have invested billions of dollars in driverless technology and tested the technology with millions of road miles.

NEW AND OLD

Many of the technologies that make truly driverless autonomous cars possible have been around for some time. However, technologies such as cameras and sensors to prevent cars getting too close to the car in front or to assist with parking have existed to assist the driver, rather than replace him or her. So not all autonomous vehicles are truly driverless, although they all use AI to make decisions.

HOW SELF-DRIVING CARS WORK

Autonomous vehicles such as the self-driving robotaxis in some cities use a wide range of sensors to gather data about their location and the environment around them:

- **Lidar** transmits millions of **laser pulses** to build a complex model of surroundings.
- **Radar** uses **electromagnetic waves** to identify objects such as other vehicles.
- Cameras gather visual data, such as traffic lights, construction signs, and street markings.
- Other sensors fix the car's position and pick up sounds such as police sirens.
- Software processes all this data to build a complete picture of the environment and so plan how the vehicle should respond. It uses machine learning to predict how vehicles and people will move.

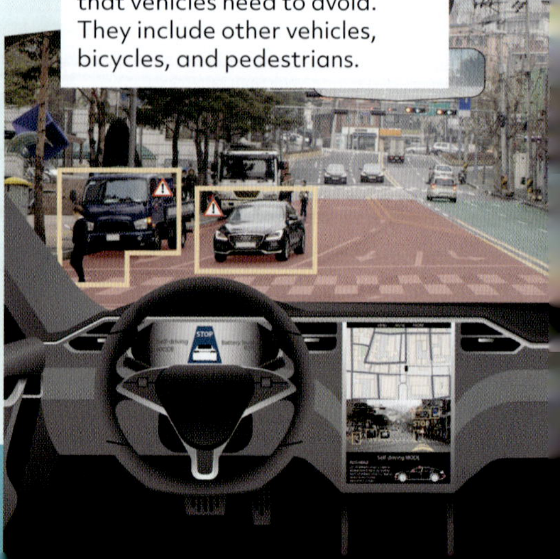

City streets are full of obstacles that vehicles need to avoid. They include other vehicles, bicycles, and pedestrians.

Self-driving cars use data from cameras, lidar, and radar to build a constantly changing map of their surroundings.

Different Levels of Autonomy

Level	Name	Description
1	Driver assistance	Some assistance with steering, **acceleration**, or **deceleration** depending on surroundings, such as if the car in front is too close.
2	Partial automation	System activates changes to steering or acceleration while the driver is there if necessary. Driver is still in charge of the vehicle.
3	Conditional automation	Automated systems perform all aspects of driving, with driver intervening if necessary.
4	High automation	Automated systems perform all aspects of driving, even if driver does not intervene. May not operate in all locations and conditions, such as bad weather.
5	Full automation	Automated systems perform all aspects of driving in all locations and conditions. Human driver not required.

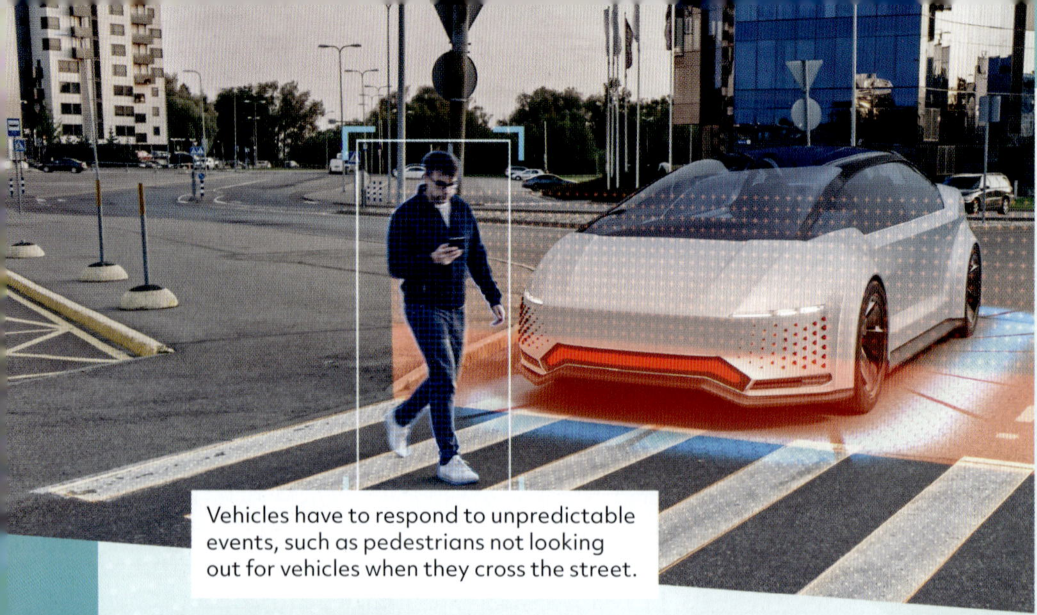

Vehicles have to respond to unpredictable events, such as pedestrians not looking out for vehicles when they cross the street.

CHALLENGES FOR DRIVERLESS CARS

Although many of the technologies used in driverless cars have existed for some time, there have been several challenges to overcome. AI systems are designed to respond to data and find solutions as a human would, but a busy street is one of the most difficult and unpredictable places to do this. Pedestrians can step out in front of cars without looking. Drivers may be distracted, in a hurry, or angry, so they ignore simple rules of the road. Sudden rainstorms or blizzards can change driving conditions and affect the information the car's computer receives through cameras.

CRASH RISK

All the hazards described above mean that the AI technology is extremely complex and a lot of driving time is required to train it in all possible problems. The risks are also high. The AI system is controlling a large steel vehicle that can cause damage and serious injury if it goes wrong. Sometimes, our computers, smartphones, and other devices crash or stop working. We normally must restart them and hope that nothing more serious has happened. This may not happen very often, but it's inconvenient. If it happens once in a car on a busy street, that's probably too often and the results could be very serious.

REPLACING THE FAMILY CAR

Driverless cars are one of the most **controversial** results of developments in AI. Cars are so important for many people that the idea of handing over control to a computer is very difficult to imagine. People sometimes even see their cars as part of their personality. Cities are built around the idea that most families own a car and it is difficult for many people to see that changing.

A WHOLE NEW WORLD

If driverless cars do start to replace the current family car, it could lead to a very different world. The cars of today are often not used as frequently as you might imagine. The car may be driven to a workplace and back, but most of the time it's in a parking lot or outside the owner's home. Many vehicles are only driven for a few minutes every day. This is becoming more common as people work from home, which could increase even more as AI changes the way that we work as well as the way that we travel.

TRANSFORMING HOW WE USE TIME

Another aspect to consider is how we could utilize driverless cars. They could transform not only how we get around, but also how we spend our time. For example, if the car can drive itself, why not send it off to pick up groceries or do the school run rather than being parked up on the drive or road? Driverless car owners could use their commuting time to work, watch TV, or even sleep. The time they once spent concentrating on traffic and controlling the car would be freed up to spend on other activities.

See pages 14–15 for more on the debate about the benefits of AI in transportation.

How would you feel if you saw a sleeping driver in a moving car? Many people feel this sight would be frightening and unnerving.

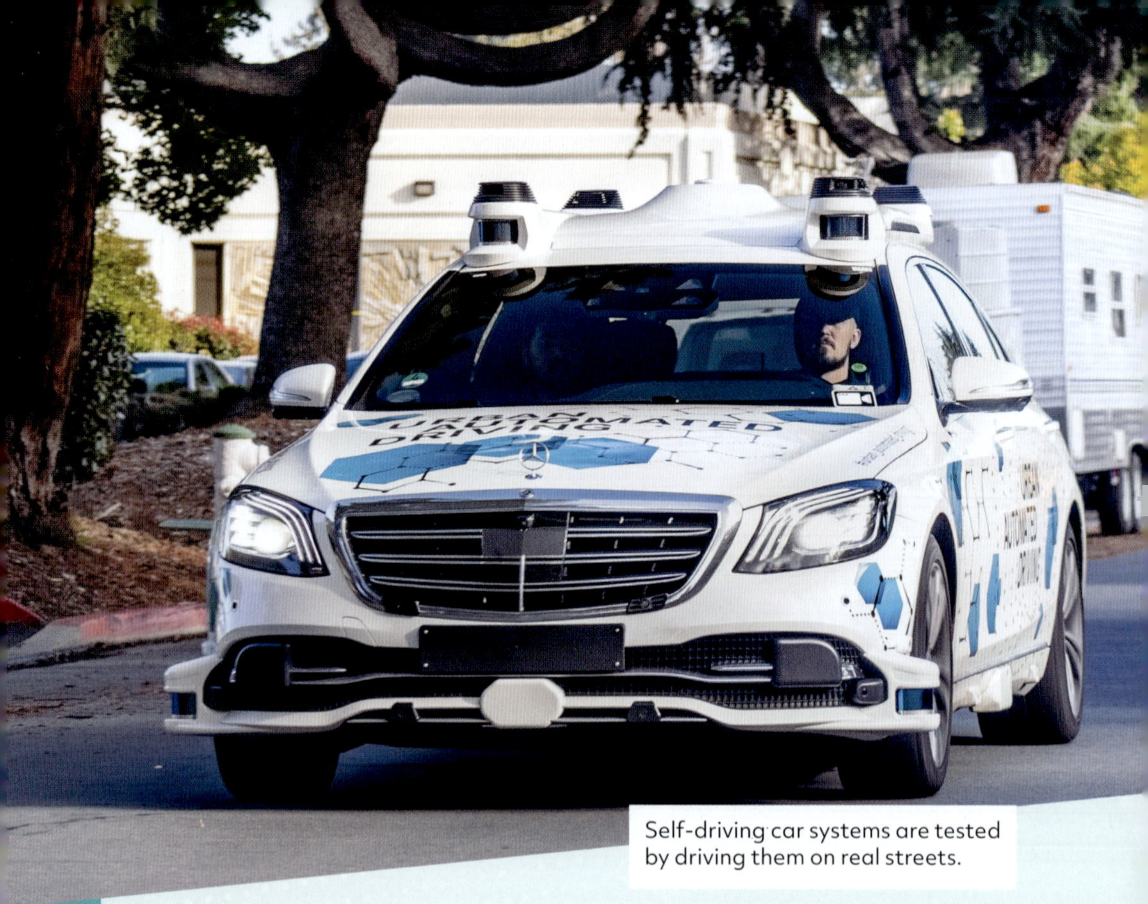

Self-driving car systems are tested by driving them on real streets.

HAIL THE ROBOTAXI!

Some people argue that if we can call up a driverless car whenever we need one, why would we need to own a car at all, especially if we live in a city? Driverless cars are powered by electricity, which means fuel costs are low and there is no cab driver to pay. Maybe we could all rely on robotaxis to get around?

ALREADY ON THE STREETS

While some of the arguments about self-driving cars are based on future developments, self-driving taxis are now in action in some of the United States' biggest and busiest cities. Some of the first taxis were launched in San Francisco, the home city of some of the biggest tech companies. San Francisco is also a tricky city to navigate, with a lot of steep and narrow streets and heavy traffic. The idea is that the AI systems controlling the cars will learn faster in an environment with more hazards and difficulties.

TAXI TROUBLE?

How did the taxis do? The main issues faced by the robotaxis seemed to be dealing with other traffic and unpredictable events, such as one car driving into wet cement. There were also examples of robotaxis getting in the way of emergency services. As the service expands to other cities, people predicted that some changes would be needed, mainly due to issues with traffic and other drivers.

DEALING WITH BREAKDOWNS

Like all cars, sometimes driverless cars break down. This creates particular problems because regular mechanics are not trained to fix software issues. Protesters in San Francisco also discovered that the robotaxis' navigation systems could be disabled by placing an orange traffic cone on the hood. The protesters were arguing that the robotaxis cause more road **congestion** and the city authorities should be investing money in improving public transportation.

JOB FEARS

One group of workers is especially concerned about the possible growth of robotaxis—cab drivers. In 2021, there were around 285,000 cab drivers in the United States. They fear that the growth of driverless taxis could take away their jobs. Cab drivers argue that their experience makes them a safer option than handing over control to a machine on busy city streets. They also believe that many people prefer a human driver to guide them and help with luggage.

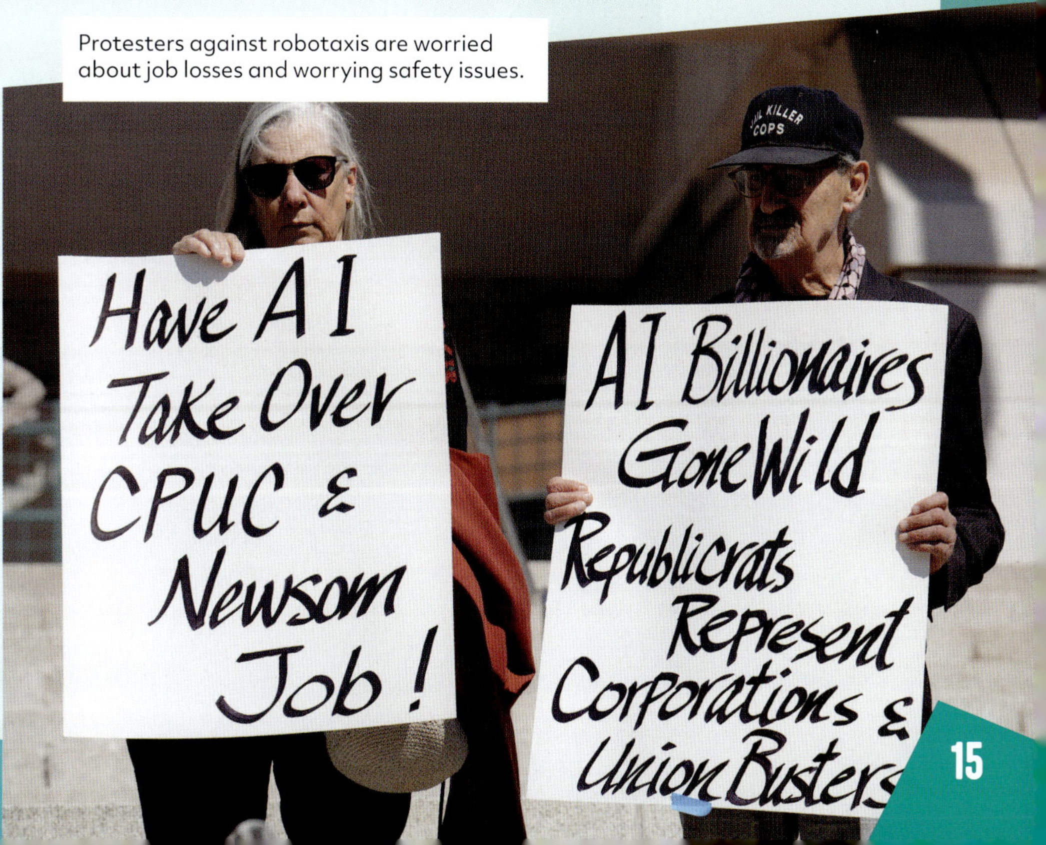

Protesters against robotaxis are worried about job losses and worrying safety issues.

TRAVEL SAFELY

Supporters of robotaxis and driverless cars in general argue that they are a safer way to travel. **Robots** don't get tired, distracted, or angry—which are all things that can make human drivers unsafe. They are also better for people traveling alone, who may not want to get in a car with a stranger.

SAVED BY ROBOTS

Any accident involving one of the driverless vehicles being tested is big news. It's worth remembering that more than 1.3 million people die in car accidents around the world every year, but people would be concerned if driverless cars were responsible for just a few deaths. Elon Musk has pointed out that the people whose lives are saved by robot vehicles don't know that their lives have been saved.

Many accidents are caused by human drivers making bad decisions or losing their temper at other drivers on the road.

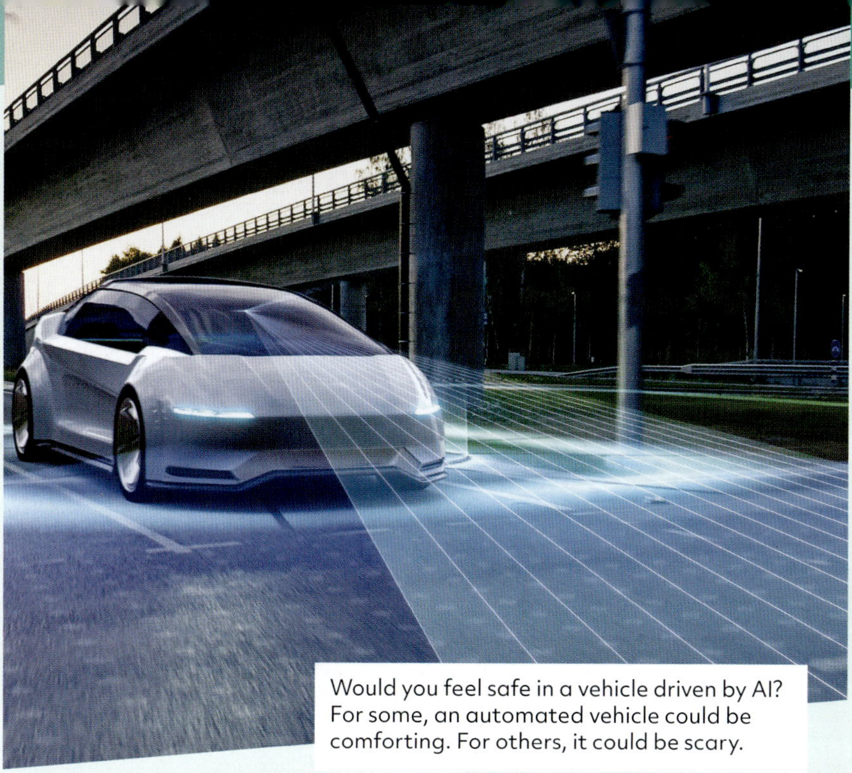

Would you feel safe in a vehicle driven by AI? For some, an automated vehicle could be comforting. For others, it could be scary.

WHAT DO PEOPLE THINK?

Whatever the claims and evidence about the safety of autonomous vehicles, there will be many people who resist any change to the way we use cars. In a 2021 survey, 73 percent of Americans said they would not feel safe in a car driven by a computer without a human driver. Surveys found a similar response in Canada and many European countries too.

FALSE STOPS

Driverless cars are designed to be as safe as possible, but this may lead to problems. The vehicle's planning software tells it when to change lanes, slow down, or stop. However, because it is designed to be cautious, objects left near or on the edge of a road, such as boxes, trash bags, or bicycles, can cause a vehicle to suddenly stop.

CARS AND CYBERCRIME

Apart from safety, other concerns could include worries about the technology used in driverless cars. Just like any computer, the complex AI software used in driverless cars could be affected by viruses and **malware**. Some people worry that the cars could be **hacked** by criminals. This could possibly lead to vehicles crashing or being redirected to a different destination than planned. This could put passengers at risk. Car makers will need to be alert and update software to deal with this risk.

The Debate:

Self-Driving Cars Will Change and Improve Travel

Some people predict that AI-controlled vehicles could be more than just another type of vehicle on the road. They say they will change and improve the way we travel altogether. There are arguments both for and against this scenario. Let's take a look at some of them.

AGREE

Less driving means longer journeys: Driverless cars will be more convenient than traditional vehicles as people will not have to concentrate on driving them. Could this mean that people will **commute** farther to work or school because time can be spent doing other things?

Different costs: Driverless car technology will make them more expensive than cars we are used to, but running costs will be lower as the cars run on electricity rather than gas. This could mean that people are happier to rent a car or use a robotaxi rather than owning one.

Changed streets: AI-controlled cars will mean fewer arguments, car horns, and pollution. There will also be less need for parking because cars can be sent home when not needed.

Connecting communities: Driverless cars will be great for people in **rural** areas or who cannot drive because they will not have to rely on public transportation. They will benefit greatly.

DISAGREE

Many people will not switch: It will be a long time before most cars on our streets are driverless. In the short term, they could just add to traffic and make congestion even worse.

New problems: Driverless cars may produce less pollution and few accidents, but they will bring new problems as technology breaks down or is hacked. This is a real and dangerous threat.

People like how they use cars now: Even if people do switch to driverless cars, they will probably use the car in the same way as they do now so any changes will be minimal.

Conclusion

As driverless cars become more common, they will certainly bring some changes to our roads, but people are often slow to change their habits and most people say they distrust AI-controlled vehicles. Changes are likely to be gradual rather than sudden and dramatic.

Q Do you think AI will help solve our transportation issues or do you think it will cause more harm?

Q Can you think of more arguments on either side of the debate?

Q Will AI be a friend or a foe? What conclusions do you draw?

CHAPTER 3

AI Transporting Goods and People

Many experts think that the first area of transportation to be dominated by driverless vehicles could be in transporting goods by road. Millions of large and small trucks deliver goods to people and businesses. AI can also help with planning how and when goods will need to be shipped using data about where they are needed. AI can do this quickly and efficiently.

AI AND CLICK TO BUY

Around the year 2000, if you wanted to buy almost anything, you would head out to the store. Online retailers such as Amazon were growing fast but goods such as groceries or clothes were not easily available online. Today, we are much more likely to order these items from the comfort of our own homes. This makes things difficult for businesses. Instead of shipping goods to stores in major cities and towns, they have to ship to thousands of different addresses. We also often expect the goods we order to arrive the next day.

AI-controlled robots are already making a big difference to the way that warehouses operate.

Driverless trucks traveling long distances could avoid some of the problems of busy city streets.

AI STEPS IN

AI has a growing role to play in making the delivery of goods happen. In the largest warehouses where your order is stored, AI systems and autonomous robots will be involved in storing and picking your order from the shelves of the warehouse. Even before this stage of the delivery process, AI has been using previous sales patterns to predict what will be ordered and where it needs to be stored.

MAPPING THE ROUTE

Making a lot of small deliveries is complex. Vehicles spend most of their time on the road between deliveries. Because the delivery locations are different each day, the quickest, most efficient route between them also changes each time. AI can help **logistics** businesses map out the route, using data on traffic flows and any other issues. This can provide a much more efficient delivery schedule, saving time and fuel.

ROBOT TRUCKS

Long-distance trucks cover huge distances, traveling for hours at a time. This leads to driver **fatigue** and a need for rest stops. Operators can save money with trucks that can operate all the time. Much of the driving time is spent on wide and straight highways, without many of the obstacles and possible issues that face driverless cars on city streets. These factors mean that there would be many advantages in these vehicles being driverless.

This driverless bus operates in Helsinki, Finland. There it is used by local people and accepted as a normal part of transportation.

SELF-DRIVING SAFETY

There are some particular safety concerns about driverless trucks. Trucks are traveling much faster than vehicles in the city. The weight of vehicles means that crashes can be very serious, especially if carrying dangerous materials. So even if the AI systems exist for the switch to driverless trucks, it is likely that drivers will remain in the cabs for a while to come.

THE LAST MILE

The most time-consuming and expensive part of any delivery journey is the last mile to the delivery location. Large online retailers will already use AI to figure out whether they need extra delivery vehicles at a particular time. In the near future, deliveries in some cities will start to be made by autonomous flying drones. They may not be able to carry heavy packages but they don't get stuck in traffic.

PLANNING PUBLIC TRANSPORTATION

Some of the AI elements that can help deliver goods more easily can also be used in planning public transportation such as buses and trains. After all, these services are designed to pick up people in one place and deliver them to another. In some ways, AI is perfect for public transportation because it is predictable—large numbers of people often take the same bus route to work or school every day. AI can easily work with that data.

INFORMATION HOTSPOTS

Public transportation systems contain a lot of data about their customers. They know how many people normally board at each train station or bus stop, where they get off, what times are busiest, and likely delays on the route. AI can also use data to figure out new routes by using data about where people live, work or go to school, and where they go to enjoy themselves.

NEW VEHICLES

Driverless vehicles traveling on rails are already a feature of some public transportation systems. We may see driverless buses and fast trains but this is probably less likely than some other vehicles. Safety is a huge concern for vehicles carrying large numbers of people, so it is likely that a driver will always be on board, even if the train or bus is controlled by AI most of the time.

AIR TRANSIT

In 2023, China went a step farther in licensing the first AI-controlled flying passenger **drones** to operate in Chinese cities. The United States is likely to not be far behind, but human pilots will probably be on board airplanes for safety reasons. AI in air transportations could ease traffic on the streets, although it may just be moving it into the sky instead, which will not help solve the ongoing issue of pollution.

Autonomous drones could carry passengers across major cities.

This driverless public transportation vehicle travels along the streets of China, picking up passengers on the way.

THE END OF PUBLIC TRANSPORTATION?

Some people think that public transportation as we know it will no longer exist for travel in towns and cities. If we can all call up a self-driving robotaxi, at low cost, and we don't have to share it, why would we choose to ride a bus? Fast trains may still be a better option for people traveling longer distances. Trains can also help avoid busy roads and traffic jams.

AN EXPENSIVE OPTION

It may be going too far to say that everyone who now catches a bus to work will soon be sitting in a comfortable driverless taxi, or that the city skies will be full of autonomous drones. After all, this would need millions of driverless cars. Although running costs are low, the cost of manufacturing these vehicles is very high, with AI, lidar, and radar systems included.

LEADING TO GRIDLOCK

Even if enough vehicles could be produced, adding thousands of driverless cars to already crowded roads would lead to traffic gridlock. There may no longer be a need to park cars on the street, but think how many cars would be needed to transport every person who currently rides a crowded bus or subway train. It would need to be an enormous number of cars. It is likely that buses will be with us for some time to come, even if AI could help make them run on time.

SOCIAL CONCERNS

One other growing concern about the possible increase in driverless vehicles in all areas is the impact this could have on our society. For people living alone, loneliness can be a serious issue. They may share a few words with the bus driver in the morning or whoever delivers their groceries. This is an important part of society and many people are worried about losing too much human contact because of AI.

SAFER TRAVEL

Many people worry about using public transportation because of safety, especially late at night. AI can improve security by keeping track of crimes such as theft. AI tools for **facial recognition** are also very accurate, so cameras on a bus or train could identify people who have not paid for a ticket, or who could be a threat to others. This should improve the experience of people using public transportation.

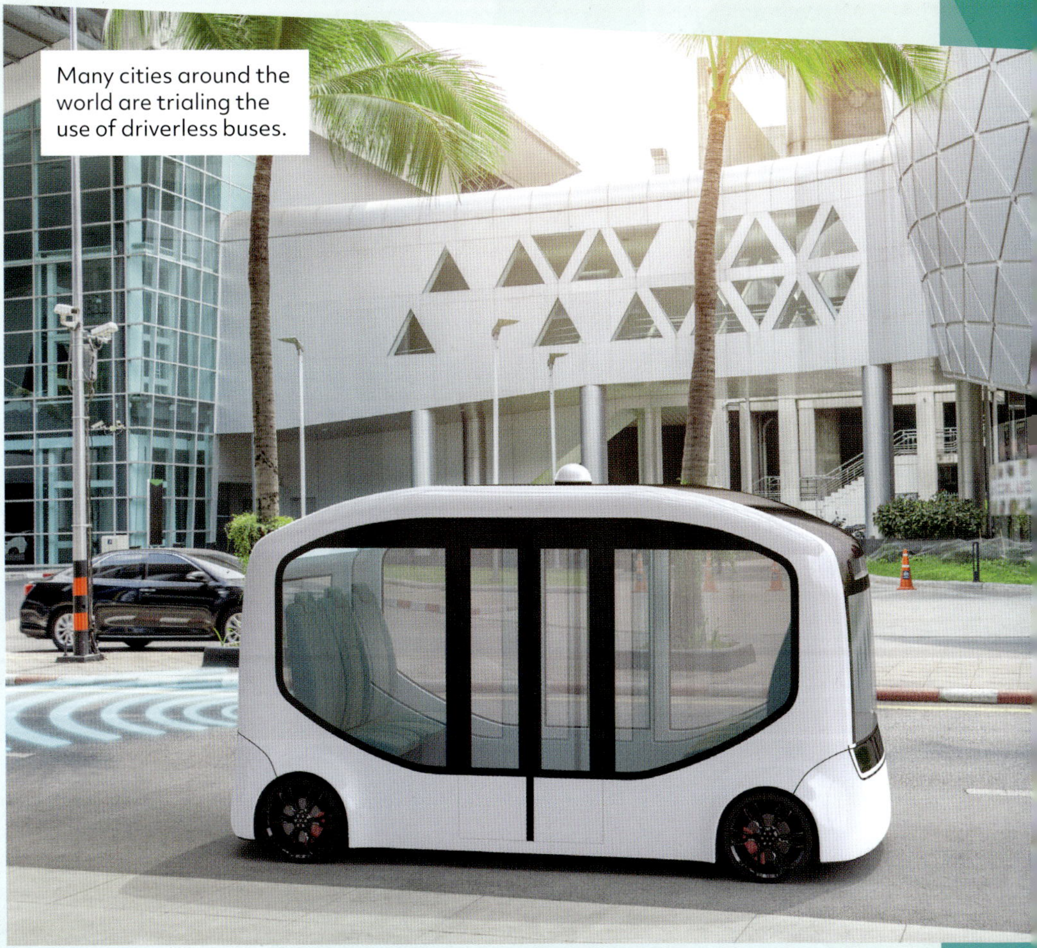

Many cities around the world are trialing the use of driverless buses.

The Debate:

AI Will End Public Transportation

Some people believe that the development of driverless taxis providing low-cost, door-to-door travel could lead to the end of public transportation in our cities. They say we will all use self-driving vehicles to get around rather than rely on public transportation. There are arguments both for and against this scenario. Let's take a look at them.

AGREE

Driverless vehicles will provide travel for all: Rather than waiting in the summer heat or winter cold for a bus or train, driverless taxis or even flying drones will take us where we want to go for a fraction of the cost we pay now, with no driver or gas to pay for. That has to be a big win.

Convenience: Driverless cars will be much more convenient—arriving when we want them to and taking us exactly where we want to go. The time spent driving could instead be spent doing something more enjoyable or productive. That has to be a big win for drivers.

Privacy and safety: We won't have to share a driverless car with other people, which will make it more private and safer, especially late at night. Many people who may feel vulnerable sharing a car with other people, especially people they do not know very well, would benefit.

DISAGREE

Established public transportation: AI may improve public transportation, including more efficient routes and safer journeys. However, we haven't proven yet that those systems will work and the transportation we currently have is a known factor for people.

Overcrowding: If everyone who takes a bus or subway train to school or work travels in a separate vehicle, this will mean even more traffic than we have now. Journeys will take too long and be less convenient. One benefit of trains is that they can avoid traffic.

Social reasons: Using public transportation is a valuable social experience for some people, whether they are older people living alone or teenagers going to school with their friends. Many would not want to lose this and AI transportation would result in that.

Conclusion

There are many good reasons why public transportation is still useful for a lot of people, although AI-controlled driverless vehicles may be a better option for some of us, we have a long way to go before they take over fully from trains and buses.

Q Do you think AI will end public transportation or do you think we still need our current systems?

Q Can you think of more arguments on either side of the debate?

Q Will AI be a friend or a foe? What conclusions do you draw?

CHAPTER 4

Improving Safety and the Environment

The developers of AI systems and driverless vehicles argue that this technology will improve our lives in several different ways. They claim that AI in transportation will make us safer than ever before and that it will be **beneficial** for society and the environment. What is the evidence for these claims?

HUMAN ERROR

One of the main reasons for the claim that AI-controlled vehicles are safer is that they reduce the chance of human errors. Some sources claim that 94 percent of all road accidents are caused by human errors. For example, 1,000 people are killed and more than 100,000 people injured every year in the United States due to drivers running red lights. Cars that use sensors and cameras to detect pedestrians and stop automatically could reduce that number significantly.

SAFER DRIVERS

Even in cars that are not driverless, AI can make a difference. Driver assistance systems can monitor if drivers are paying attention to the road and warn them of any issues. One major AI development has already made roads safer—**voice-recognition** software. Now drivers can communicate by phone or even give instruction to the vehicle without taking their eyes off the road for an instant. No one knows how many lives could be saved by this AI innovation that is available to anyone with a smartphone.

Accidents often happen when drivers have to deal with a lot of different demands, such as at busy city intersections.

Some people question whether AI will be able to deal with unpredictable conditions, such as snow-covered roads.

AI INSPECTION

Maintenance is another major safety issue. Inspecting cars for defects that might cause accidents is time-consuming and costly. Now AI systems can use cameras to give a visual inspection of a car in seconds. This can identify likely issues, such as tires that may not have enough grip. This helpful AI tool will make all vehicles safer.

SMART ROADS

The benefits of AI are not limited to the vehicle and its driver. AI is also making roads safer. AI can use the data from thousands of cameras and other data to improve safety and efficiency. For example, road sensors can measure the speed and weight of vehicles and smart traffic lights collect information about traffic at busy intersections.

MONITORING WEATHER AND TRAFFIC

AI systems use the data they have gathered about the vehicles on the road along with other information, such as weather conditions or road maintenance. They can then adjust systems such as traffic signals or speed limits in bad weather to make traffic flow more easily and safely. The data gathered by smart roads can also be used to identify traffic offenders and help emergency services to reach emergencies far more quickly.

ENVIRONMENTAL IMPACT

Transportation has a major impact on the environment. This includes **noise pollution** and fumes that can have serious effects on people's health. Non-electric vehicles also release carbon dioxide into Earth's **atmosphere**. This is one of the main causes of climate change. Tackling the harmful environmental effects of transportation is very important, but how can AI help?

CLEANER CARS

The breakthroughs in driverless cars are happening alongside a switch to electric vehicles. All driverless cars are electric, which means they are not producing carbon dioxide in the same way as older cars. They are also quieter than traditional cars.

CONTROLLING ENERGY USE

AI control has a direct effect on how much energy driverless vehicles use. AI will drive a car at a constant speed, which can reduce energy use by 15 to 20 percent. Speeding up and slowing down suddenly uses more energy and so creates more pollution. AI will stop this problem.

DOWNSIDES OF DRIVERLESS

It's not all good news with driverless cars. Manufacturing new cars uses energy. Battery technology that is essential for electric cars uses some rare materials which are only found in quite small quantities. Powering powerful computers, radar systems, and cameras needed for AI-controlled cars also uses a

Electric vehicles do not produce engine fumes but a lot of energy is used to **manufacture** and run them.

Energy use could be higher because driverless cars spend more time on the road.

lot of energy. In fact, the systems needed to control cars by AI could actually use more energy than cars powered by gasoline. That's one reason why driverless cars are all electric—because they would use up too much fuel if they were not.

FEWER CARS

There are also reasons why driverless cars could mean we need fewer cars on the road. If a driverless car can make multiple trips to take different family members to work or school, or to pick up groceries, families will not need to own multiple cars. City dwellers who rarely use a car may not need to own one at all; calling a robotaxi when they need a car.

LONGER JOURNEYS

But the introduction of AI could also mean changes in habits, with AI-controlled cars in use for much more of the time as people send them home rather than parking in the city. Commutes could become longer as people are happy to spend more time in the car as passengers than they did as drivers. Research has found that drivers of partially automated vehicles do make longer journeys because of the transformative technology.

See pages 34–45 for more on the debate about the advantages of AI in transportation.

Cutting congestion could make journeys quicker and cause less damage to the environment.

CUTTING TRAFFIC

AI can also make transportation more efficient. Cars that are stuck in traffic are bad for the environment if their engines are still running. That's where smart roads and traffic signals can help. In the city of Seattle alone, 1,000 tons (907 mt) of carbon dioxide are pumped into the atmosphere by cars waiting at red lights every single day. AI is being used to reduce these wait times by a few seconds each time, which could cut emissions by 10 percent.

ECO-FRIENDLY ROUTES

In 2021, Google introduced key changes to its mapping software that use AI to try and reduce carbon dioxide emissions. As a result, every time users requested directions, Google's mapping software would not just show them the quickest and shortest route, but also one that would use the least fuel. This could be a route with less traffic or fewer intersections where the vehicle would have to stop. Complex AI is needed to crunch that data and give an instant response. It is believed the feature could save 1 million tons (approx. 900,000 mt) of carbon emissions each year. It could also help drivers save on the cost of fuel.

TAKING SAFETY FOR GRANTED

One worry about the increased safety features in partly or fully autonomous cars is that we will come to rely too much on AI. Drivers may not pay attention as they did in the past, relying on the cars cameras and sensors to pick up any problems. In fully driverless vehicles, we may not know how to step in if something goes wrong.

Future generations of car users may not even learn to drive themselves so they will be completely reliant on AI to get them safely from one place to another. Is that sensible?

DATA AND PRIVACY

Many of the environmental and safely benefits discussed in this chapter involve AI systems using data in ways that was not possible until recently. For example, human planners knew that vehicles use a lot of fuel waiting at traffic lights, but figuring out how to improve this using up-to-the-minute data would have been impossible without AI. AI-controlled vehicles do collect huge amounts of data, which is needed to make them work. Analyzing this data could tell companies and others a lot about us, including where we travel to and from. There may need to be more rules about how this data can be used, to protect our privacy.

AI systems can use data to make connections and understand how we make decisions.

The Debate:

AI Will Improve Our Roads

Driverless electric cars will reduce pollution in our cities, and many people hope that AI itself will be able to make travel safer and cleaner. There are arguments both for and against this scenario. Let's take a look at them.

AGREE

Removing human error: A large proportion of road accidents are caused by drivers making mistakes or being distracted. AI using cameras and sensors should make that far less of a problem.

Fewer cars: Driverless cars will be more efficient and able to make many journeys every day to take people to work or school. This will mean there are fewer cars on the road, using less energy and making roads less crowded. That surely has to be a key benefit.

Smart roads and traffic signals: AI can use data about the way we travel to make traffic flow more easily and reduce time spent in jams. This will mean a big reduction in carbon dioxide emissions and pollution. Again, that's another big win for AI.

DISAGREE

Safety issues: Cutting down on human error is important, but AI-controlled cars could bring new risks as they fail to deal with unexpected problems. They will also have to use the same roads as human drivers and other people making mistakes.

Longer journeys: Not having to drive themselves may persuade people to live farther from where they work or travel farther at other times, so roads may get even busier. That is a key concern.

Switching from public transportation: Convenience of driverless cars could make people switch from public transportation, which will mean more cars on the road than before leading to more traffic congestion and accidents, even if the cars themselves are cleaner.

Conclusion

Safety is one of the biggest reasons given by supporters of driverless cars. However, there are ongoing concerns about how safe these cars are in all locations. AI should certainly help reduce carbon emissions and pollution, which is a big benefit, but driverless cars still use a lot of energy and other **resources**, so they are not without problems.

Q Do you think AI will help improve our roads or do you think the tech is too problematic?

Q Can you think of more arguments on either side of the debate?

Q Will AI be a friend or a foe? What conclusions do you draw?

CHAPTER 5

AI in the Sky

You might think that creating aircraft that could fly using AI rather than human control would be more difficult than designing self-driving cars. There is one big issue that AI-controlled aircraft don't have to deal with: there are fewer hazards to avoid in the sky such as slow-moving pedestrians or traffic lights. However, there are also several challenges to the development of AI in the sky.

DRIVERLESS DRONES

Unmanned aircraft are called drones. They are often flown by remote pilots on the ground and are used extensively by the military. However there are some examples of autonomous drones piloted by AI systems. In 2023, a small autonomous drone controlled by AI was able to beat the best human drone operators in a race round a complex course, dodging obstacles at speeds of around 100 miles per hour (160 kph). Although these drones were lightweight, they showed that AI could navigate a difficult course and make very quick decisions.

Flying taxis could be the future of city transportation.

As AI takes more control, the pilot's role may become more about monitoring the systems and taking control only in certain situations.

REAL-WORLD USES

In the real world, autonomous drones are used to inspect structures that are difficult for people to reach, such as bridges, offshore oil platforms, and wind turbines. Cameras and other sensors can automatically record any maintenance issues such as cracks in the structure. Autonomous drones are likely to become a more common sight, delivering parcels and even transporting people around cities in the near future.

PILOTLESS AIRPLANES

When you board an aircraft, are you reassured by the calm voice of the captain telling you that everything is going well on the flightdeck and the weather is fine on the ground? Maybe you wonder who's flying the plane while they're talking to you? Pilots are skilled and highly trained individuals, but on modern aircraft much of the work of flying the aircraft is done by autopilots and other automated systems using AI. The pilot spends much of their time monitoring these systems —ready to take control if something does go wrong.

SAFER AIR TRAVEL

There is optimism that AI can make air travel safer. Almost every single aspect of an aircraft is recorded electronically, creating more data than humans are able to monitor easily. Each aircraft **engineer** also makes notes about the aircraft and how it is maintained. AI can use data about an aircraft's performance to predict which parts will need replacing and which parts of the aircraft are not working well. This constant monitoring can reduce the amount of time an aircraft needs to spend on the ground, as well as reducing the chances of an emergency in the air. These benefits of AI in air travel are persuasive arguments in its favor.

AI can learn from how pilots react to difficult weather conditions. It can then build that information into its own responses too.

EMERGENCY AI CONTROL

AI researchers are now developing autopilots that can learn from what the best pilots do in emergency situations, such as flying in bad weather or reacting when an aircraft's systems fail. Human pilots may be affected by stress or poor training, so AI autopilots will be able to make human errors less likely.

AUTONOMOUS AIRCRAFT

However, air travel experts predict that fully autonomous passenger aircraft may not be too far away. Autonomous cargo flights monitored by people on the ground are already being planned. Of course, an airplane full of passengers is a different matter. It's difficult to imagine that any passenger flight would not have a trained pilot on board, but their role might simply be there as a backup for the aircraft's AI systems.

AI EFFICIENCY

Just as on the roads, AI in the air can make air travel run more efficiently. There may be fewer obstacles to consider but planners still need to think about weather patterns and keeping distance between air traffic to avoid any chance of a collision too. Alaska Airlines started using an AI tool to plan flights. It was able to schedule flights to avoid bad weather or busy landing slots. The smart system saved about 480,000 gallons (1,816,000 l) of fuel in 6 months and cut flight times. AI could also make things easier for passengers, such as using facial recognition to check identities and speed up boarding.

TOO RELIANT?

Governments and air travel organizations are rightly cautious about giving too much control to AI. In 2018 and 2019, two Boeing 737 Max aircraft crashed in Indonesia, Asia, and Ethiopia, Africa. It was later found that the crashes were caused by flaws in a piece of onboard software. The software was designed to support pilots but instead put the aircraft into a dive that the pilots could not correct. Many lives were lost. This shows the danger of relying on sophisticated AI systems that may contain hidden problems.

Training AI Autopilots

AI systems to pilot an aircraft need to be trained just like any human pilot. The process is called supervised machine learning. A human pilot shows the system how to behave and what to do in certain situations. This data can be used to train the autopilot and build up a body of data for future training.

Air traffic control can be made more efficient and safer by using AI tools.

The Debate:

Aircraft Will Be Piloted by AI

Aircraft were some of the first vehicles to be controlled by automatic systems and AI systems are already involved in controlling most large aircraft. Perhaps one day all aircraft will be piloted by AI. There are arguments both for and against this scenario. Let's take a look at some of them in this debate.

AGREE

AI removes human error: There are far fewer aircraft accidents than there are on the roads, but human error is a factor in many of them, whether they are caused by the pilot or are the result of a maintenance error that has not been picked up by engineers.

Aircraft may be easier to control than road vehicles: While aircraft are very large and complex machines, they do not have to dodge pedestrians or road vehicles. They also don't need to stop at intersections. This makes them more predictable for AI to control.

AI is controlling many aircraft systems: AI is already piloting the aircraft in many situations, so maybe it's not such a big leap for AI-controlled aircraft systems to take over from pilots completely.

DISAGREE

Pilots are highly trained: Almost anyone can drive a car, if they've passed a driving test. Pilots have many years of training in how to deal with every situation so aircraft should always have a pilot.

Risks are much greater than other vehicles: If something goes wrong in an aircraft carrying many passengers, the risks are much greater than in a motor vehicle. An aircraft can't just stop its engine in mid air so that the passengers on board can safely escape.

Computer systems can fail: There are many examples of computer systems failing or not functioning as they should. It is essential that a pilot is able to take control in an emergency.

Conclusion

AI certainly has a part to play in air travel, making it safer and more efficient. Most flying may even be controlled by AI. Small aircraft or those carrying cargo may be able to fly without a pilot, but the safety risks are too great for this to happen with larger passenger aircraft.

Q Do you think AI can be used to fly all of our airplanes or do you think it is too dangerous?

Q Can you think of more arguments on either side of the debate?

Q Will AI be a friend or a foe? What conclusions do you draw?

CHAPTER 6

Friend or Foe?

The many uses for AI in transportation are starting to appear in our cities, from driverless robotaxis to hidden AI tools that make travel safer and more efficient. In the long term, AI could lead to huge changes in the way we travel. People are very excited about the benefits of AI but cautious about some of the possible negative effects on our lives. There will be winners and losers from the AI revolution.

WHO WILL BE THE WINNERS?

Driverless cars could lead to a lot of positive benefits for travelers. Many people will welcome the time-saving benefits of not driving themselves. AI-controlled cars should bring better road safety and environmental benefits too. Many people predict that the cost of new robotaxis will be much lower, which will be a big benefit for people who are unable to drive or don't have access to public transportation.

Driverless cars should be a good thing for crowded cities too. There will be less pollution and noise from electric vehicles. Less space will be taken up by parking lots and cars parking on the street. There may even be fewer cars on the road altogether. AI will also help make public transportation more efficient.

Changes in how we travel will mean major changes in the car industry. Some of the leading makers of driverless cars are technology companies rather than traditional carmakers. Some countries in particular, such as China, are putting a lot of money into developing AI technology.

WHO WILL BE THE LOSERS?

In many industries, people fear that AI will lead to job losses, and transportation is no different. There are millions of people who work as truck drivers in the United States. There are also many others who work as taxi and bus drivers and numerous other driving jobs. If trucks and taxis are driven by AI, there will be far fewer jobs available in these industries. The human impact of AI technology in transportation is a major concern for many people opposed to it.

There are also a great number of other businesses that will be affected by the growth of AI in transportation, from driving schools to parking lots. Traditional carmakers may find that they are left behind by the AI systems of smart technology companies.

If we are to switch to driverless cars, there will have to be some trade-offs. The complex technology in these cars could make them too expensive for many car drivers. There are also plenty of people who don't want to switch to driverless cars or AI-controlled drones. Many of these people don't trust a computer to control their car.

The possible benefits of AI in transportation are huge, but there are also problems to be solved by engineers, drivers, and governments. In 2023, one robotaxi provider had to stop operating because of issues around the way its technology reacted to an accident. The company was accused of not revealing important information.

As we have learned, the road to a fully automated transportation system may not be entirely smooth. However, the key benefits of AI are clear and the technology could help us solve many of our ongoing problems with transportation.

Will driverless cars make our cities cleaner and less congested?

The Ethics of AI

To ensure that AI's impact on transportation is positive, developers and governments need to think about the **ethics** in AI development. The important questions about future transportation outlined below need to be asked.

- How important is safety? We know that use of AI can make transportation safer, but much will depend on **algorithms** and the way AI is trained to make this happen.
- How does AI respond to an accident or unexpected situation? When drivers have to react to an emergency, they may make instant decisions about how to do so. How will AI make these decisions?
- How much control should we give up to AI? If AI is found to be safer than human drivers, should it be allowed to take over completely?
- Do cars need licensed drivers? Should an AI-controlled car be able to take children to school without an adult driver being present? This raises questions about who is to blame if something goes wrong.

Q What rules would you put in place to ensure that AI results in positive changes for everyone?

Q Would you add anything to this list?

Find Out More

BOOKS

Fallon, Michael. *Self-Driving Cars: The New Way Forward*. Lerner Publishing Group, 2019.

Gitlin, Martin. *Smart Vehicles* (21st Century Skills Innovation Library: Exploring the Internet of Things). Cherry Lake, 2020.

Jackson, Tom. *Robots, Gadgets, and Artificial Intelligence* (Very Short Introductions for Curious Young Minds). Oxford University Press, 2022.

ONLINE

How do AIs such as ChatGPT learn? This video explains all:
https://youtu.be/R9OHn5ZF4Uo

Visit ABC News to discover the documentary of a TV journalist who takes a wild ride inside Waymo's self-driving car in San Francisco (it does not all go to plan!):
https://youtu.be/-RxvI3INKSg

You can discover more about driverless cars on websites of manufacturers such as Waymo (waymo.com) and Tesla (tesla.com).

To find out more about AI you can also search for websites of companies involved in AI such as OpenAI, the creator of ChatGPT, and Google's DeepMind.

Publisher's note to educators and parents:
All the websites featured above have been carefully reviewed to ensure that they are suitable for students. However, many websites change often, and we cannot guarantee that a site's future contents will continue to meet our high standards of educational value. Please be advised that students should be closely monitored whenever they access the Internet.

Glossary

acceleration speeding up

algorithms processes or sets of rules to be followed for a computer to solve a problem or perform a task

artificial describes something made by humans and not naturally occurring

atmosphere the layer of gases that surround a planet

autonomous able to operate without human control

autopilots systems than can pilot a vehicle with little or no human involvement

beneficial bringing benefit

carbon dioxide a gas emitted when fossil fuels are burned, which is a major cause of climate change

climate change the gradual change in weather and temperature over a long period of time

commute a journey that someone makes regularly to work or school

congestion too much traffic in a particular area

controversial causes debate and concern

debate an argument or discussion about a particular subject, in which arguments are given for and against the main question

deceleration slowing down

drones powered aircraft without human pilots on board

electromagnetic waves waves of energy that are created because of vibrations between an electric field and a magnetic field

emissions substances that are produced or discharged, such as when fossil fuels are burned

engineer a person who designs and builds machines and structures, including computer software engineers

ethics moral rules that decide how people behave

facial recognition describes software that can recognize faces of individuals, such as when the user's face unlocks a smartphone

fatigue extreme tiredness

foe an enemy

fossil fuels fuels such as coal and oil that are made from the remains of prehistoric living organisms

gyroscope a device used in autopilots and compasses to determine when something moves from its planned course

hacked broke into a computer system

laser pulses pulses of intense light

lidar using lasers to detect objects in the area around something

logistics managing how objects are stored and moved from one place to another

machine learning the process by which computers can learn and adapt without clear instruction by analyzing patterns in data

maintenance the process of looking after and repairing something

malware software that is designed to disrupt or damage a computer system

manufacture to make products from materials or different parts

navigate to determine a location and plan directions

noise pollution excessive noise in the environment

pedestrians people who are walking rather than traveling by vehicle

pioneers the first people to do something

pollution harmful substances or other things introduced to the environment, which can include fumes or hazardous materials but also excess noise

predict to say what will happen in the future

primitive very basic and simple

radar using radio waves to detect objects in the surrounding area

resources things that people need and use, such as fuel

robotaxi an electric taxi controlled by AI that operates without a driver

robots machines that can replicate human functions or movements automatically

rural describes an area in the countryside, not in a town or city

software programs or instructions that affect how a computer operates

unpredictable difficult to predict or plan for

voice-recognition able to recognize and respond to the human voice

Index

About the Author

Nick Hunter is a highly experienced children's book author, who has written countless titles on many subjects, from history and science through social studies and geography. In writing this book he has discovered that AI is an incredibly powerful technology that has the potential to bring great benefits to transportation if we manage its potential risks.